I0475123

NOT DENUMERABILITY OF RATIONAL NUMBERS

NOT DENUMERABILITY OF RATIONAL NUMBERS

ROBERTO MUSMECI

Publisher: Roberto Musmeci

June 2014

iii

First Printing: June 2014

ISBN–13: 978–1–312–24638–6

Publisher:
Roberto Musmeci
Via Adua, 86
Monterosso Almo, Italy 97010

Contents

1   Introduction

In 1874, Georg Cantor published a paper containing the analysis which has been considered a demonstration of the denumerability of rational numbers. Even though the demonstration did not employ a rigorous mathematical formalism, it was accepted without any further development apart from its rendering as a table in the well-known form. This paper shows that the theorem of Cantor is erroneous in its demonstration as well as in its conclusions. It is therefore also a critique of the use of non-formal methods in mathematical demonstrations. The concept of denumerability of a set is, like that of infinity, based on the concept of natural numbers, the rigorous definition of which was given by the Italian mathematician Giuseppe Peano. Usually any set is considered denumerable if it can be put in a one-to-one correspondence with Natural Numbers. Although this definition is acceptable for finite sets, it cannot be applied to infinite ones. There are people who consider the possibility of finding a correspondent of one element, chosen at random from the first set, in the second set and vice versa acceptable as a demonstration. This method must be rejected however, since the possibility of choosing one number from the first set and another from the second, without their being linked by a functional law, does not guarantee that we are not dealing with a subset of one of the two sets. In the first section of this paper, I argue that this is exactly what happens in the currently accepted demonstration of the denumerability of rational numbers. The impossibility of putting the sets of rational and natural numbers in one to one correspondence is also demonstrated. In the second section, a formalization of the definition of denumerability is proposed, based on the Peano's definition of Natural Numbers. Compared with the generally accepted definition, this new definition has the advantage of being operative, i.e. it allows us to decide (or prove) if a set is denumerable or not by verifying whether it satisfies the conditions in the definition of denumerability.

2   The original Cantor's demonstration

In the article "Über eine Eigenschaft des Inbegriffes aller reellen algebraischen Zahlen" Cantor demonstrated that the algebraic real numbers are denumerable, that is, they are in a one-to-one correspondence with the Naturals, ordering them on the basis of the highness N:

> If we turn to equation 1 [cfr. $a_0\omega^n + a_1\omega^{n-1} + \ldots + a_n$], satisfied by an algebraic number ω and fully determined thanks to the conditions we have imposed, we can call *height* of ω and indicate with N the sum of the absolute values of its coefficients plus $n-1$ where n is the grade of ω; it will therefore be, applying a now usual notation,
>
> $$N = n - 1 + |a_0| + |a_1| + \ldots + |a_n|. \tag{2.1}$$
>
> The height N will therefore be a positive natural number determined for every algebraic real ω; on the other hand, for every positive natural value of N there exists only a finite number of algebraic real numbers of height N. Let $\phi(N)$ be this number; then it will be, for example, $\phi(1) = 1$, $\phi(2) = 2$, $\phi(3) = 4$. Thus it is possible to order the elements of the class (ω), that is all the algebraic real numbers, in the following manner: put in first place, as ω_1, the single number of height N = 1; this will be followed, in increasing order, by the $\phi(2) = 2$ algebraic real numbers of height $N = 2$, indicated by ω_2, ω_3; then, in increasing order, there will be the $\phi(3) = 4$ numbers of height $N = 3$; and in general after having, in this way, numbered and collocated in a determined position all the elements of ω up to a certain height $N = N_1$, we will follow them, always in increasing order, with the algebraic real numbers with height $N = N_1 + 1$. In this way we obtain the class ω of all the algebraic real numbers in

the form

$\omega_1, \omega_2, \ldots, \omega_\nu$

and with this ordering, we can speak (without any element of class ω being omitted) of the ν-th algebraic real number.

Although the argument of these pages concerns Rational Numbers, it is opportune to highlight one point, which will be amplified below, regarding Cantor's demonstration. In the demonstration cited, Cantor never uses the concept of infinite or that of limit. He implicitly uses the principle of induction which is a fundamental part of the definition of Natural Numbers. He then makes a leap in logic by claiming to obtain all the Algebraic Real Numbers. This leap is not justified (or demonstrated) and is taken for granted. Given that what must be demonstrated is the nature of the set of Algebraic Real Numbers, the use of the term all, which implies their denumerability, within a demonstration which aims to prove this denumerability, makes the demonstration itself self-referencing and thus invalid. The term all[1] implies a passage to single or double limits which must be justified. As will be shown in this paper, the passage is necessarily to double limits and cannot be reduced to a single limit, so that the term all in Cantor's paper is not only unjustified but also wrong. A more accurate and mathematical analysis of the problem of the denumerability of rational numbers is therefore required.

---

[1] As can easily be seen, as the argument of ϕ gradually increases, the value of $\phi(\nu)$ also increases, i.e. in mathematical symbols: for $\nu \to \infty$, $\phi(\nu) \to \infty$. On the other hand, for $\nu \to \infty$ not only the value of $\phi(\nu)$ but also the number of $\phi(\nu)$ tend to infinity. Thus it is no longer so evident that it is possible to put all the algebraic real numbers in bi-univocal correspondence with $\aleph$.

3 The tabular Demonstration

In order to make the discussion clearer, I report below one of the formulations of the theorem of the denumerability of rational numbers, derived from Cantor's theorem, taken from "Elements of theory of functions and functional analysis" by Kolmogorov and Fomin [3, pp.5–6]:

Theorem 3.1. *The sum of an arbitrary finite or denumerable set of denumerable sets is again a finite or denumerable set.*

Proof. *Let $A_1, A_2, \ldots$ be denumerable sets. All their elements can be written in the form of the following infinite table:*

$$
\begin{array}{cccccc}
a_{11} & a_{12} & a_{13} & a_{14} & \ldots \\
\\
a_{21} & a_{22} & a_{23} & a_{24} & \ldots \\
\\
a_{31} & a_{32} & a_{33} & a_{34} & \ldots \\
\\
a_{41} & a_{42} & a_{43} & a_{44} & \ldots \\
\\
\ldots & \ldots & \ldots & \ldots & \ldots
\end{array}
\tag{3.1}
$$

where the elements of the set A_1 are listed in the first row, the elements of A_2 are listed in the second row, and so on. We now enumerate all these elements by the "diagonal method", i.e. we take a_{11} for the first element, a_{12} for the second, a_{21} for the third, and so forth, taking the elements in the order indicated by the arrows in the *following table:*

$$
\begin{array}{cccccc}
a_{11} & \rightarrow & a_{12} & a_{13} & \rightarrow & a_{14} & \cdots \\
& \swarrow & & \nearrow & & \swarrow & \\
a_{21} & & a_{22} & a_{23} & & a_{24} & \cdots \\
\downarrow & \nearrow & & \swarrow & & & \\
a_{31} & & a_{32} & a_{33} & & a_{34} & \cdots \\
& \swarrow & & & & & \\
a_{41} & & a_{42} & a_{43} & & a_{44} & \cdots \\
& & & & & & \\
\cdots & & \cdots & \cdots & & \cdots & \cdots
\end{array}
\qquad (3.2)
$$

It is clear that in this enumeration every element of each of the sets A_i receives a definite index, i.e. we shall have established a one-to-one correspondence between all the elements of all the $A_1, A_2, \ldots$ and the set of natural numbers. This completes the proof of our assertion.

4　Some informal considerations

A set is conceived as denumerable if it is possible to count the elements without missing any, at least in principal. The count must be exhaustive and, therefore, cannot depend on the order in which the elements to be counted are placed. In the matrix arrangement it is clear that, without resorting to Cantor's expedient, it is not possible to count all the elements of the doubly infinite matrix. In fact, each line (and also every column) is infinite, even if denumerable, and it is impossible (by definition of denumerable infinite) to complete its numeration. On the other hand, Cantor's arrangement seems to give the result of making denumerable the doubly infinite matrix. "Seems" because it is possible to "interpret" Cantor's arrangement in a very different way. I rebuild the demonstration of Cantor below, evidencing the property of symmetry contained in it. Beginning with the first diagonal:

$$a_{11} \tag{4.1}$$

continuing with the second:

$$
\begin{array}{ccc}
a_{11} & \rightarrow & a_{12} \\
 & \swarrow & \\
a_{21} & & a_{22}
\end{array}
\tag{4.2}
$$

then the third:

$$
\begin{array}{ccc}
a_{11} & \rightarrow & a_{12} & & a_{13} \\
 & \swarrow & & \nearrow & \\
a_{21} & & a_{22} & & a_{23} \\
\downarrow & \nearrow & & & \\
a_{31} & & a_{32} & & a_{33}
\end{array}
\tag{4.3}
$$

etc.

At each step we add the elements of the diagonal d_N (whose members a_{ij} are identified by the property that $i + j = N + 1$). But to each counted diagonal d_n with $n < N$, there corresponds a diagonal d_m with $m = N + (N - n) = 2N - n$ not yet listed. Proceeding to infinity we find that, if on one hand it seems logical that any element of the table is reached and numbered, on the other the mathematics of the indices shows that an infinite number of these will be ignored, $N \to \infty$, $(2N - n) \to \infty$, $\forall n$. Which of these two demonstrations is to be considered correct? Our thesis is that neither of them is valid. What the above argument in fact demonstrates is that it is not possible to base a demonstration on a non-rigorous language, especially when handling abstract concepts such as that of infinite. An appeal to evidence in mathematics risks concealing something which is preconceived and which we are not able to demonstrate in the true sense of the term, that is, using the data and procedures of the theory in the context in which we are formulating the theorem. What's happened then? Happened that a dimension was hidden by the special arrangement of the table. As an example let us consider a square; if we think about its properties you will surely agree that it is a two-dimensional geometric element with two bounds of the same width.

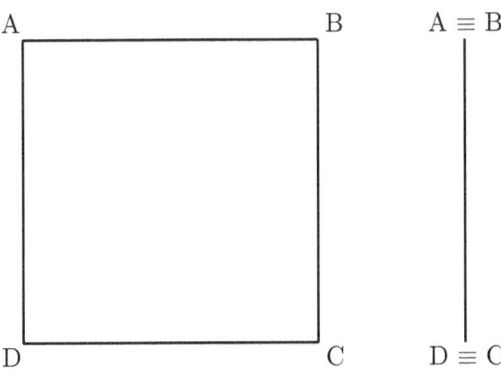

If someone turn it until you will be faced to one of its faces and will ask you about what kind of object it is now, you will surely answer that it is still a square with a side towards you. The special direction from where you see it doesn't modify its properties. That's surely true, and I think that this is a generally valid statement. So this is another, and really striking, logical deduction that the reasoning of Cantor cannot be right. The matrix of numbers of the proof of Cantor's theorem cannot be numbered row by row (and also column by column, of course) because each row is infinite. And so it must be whatever the arrangement. If it seem denumerable, it only seems! As already said, to solve a mathematical problem, it is necessary to use th mathematical formalism and a rigorous mathematical proceeding. In the following paragraphs it will be analyzed the Cantor's theorem proof and it will be shown that it is not correct. Next a new definition of denumerability will be proposed. The new definition is founded upon the definition of Natural Number formulated by Peano. It has the advantage to introduce a method to decide if an infinite set is denumerable or not based only upon the properties of the set.

5   The rigorous analysis

Let us return to the previous theorem and rigorously analyze the demonstration. Let A_1, A_2, ... be the sets of the theorem verifying the same conditions, i.e. that they are non-intersecting two by two. Call $a_{1\ 1}$, $a_{1\ 2}$, $a_{1\ 3}$, ..., the elements of A_1; $a_{2\ 1}$, $a_{2\ 2}$, $a_{2\ 3}$, ..., the elements of A_2; $a_{n\ 1}$, $a_{n\ 2}$, $a_{n\ 3}$, ..., the elements of A_n; etc. The sets A_n form a succession (a succession is denumerable by definition) $\{A_n\}$ with $n \in \aleph$. Also the elements of the sets A_n form successions $\{a_m\}_n$, $m \in \aleph$, $\forall n \in \aleph$. If we attempt to form a succession putting the elements of A_1 in the first row, then those of A_2 and so on, we will only obtain a repositioning, purely informal in that it is not exhaustive, of the elements of A_n: A_1; $a_{2\ 1}$, $a_{2\ 2}$, $a_{2\ 3}$, ..., $a_{2\ 1}$, $a_{2\ 2}$, $a_{2\ 3}$, ..., $a_{n\ 1}$, $a_{n\ 2}$, $a_{n\ 3}$, ... Cantor's theorem claims that by modifying the order in which the elements of the various A_n are ordered it is possible to number all of them. If the new order is described rigorously we can see if it is possible to put it in bijective correspondence with $\aleph$ or not (see table 3, pag. 5). If we consider the elements of the diagonal and call D_1 the set formed by the elements of the $i - th$ diagonal, we have:

$$D_1 \equiv \{a_{11}\};$$
$$D_2 \equiv \{a_{12}, a_{21}\}; \ldots ; \qquad (5.1)$$
$$D_n \equiv \{a_{1\ n}, a_{2\ n-1}, \ldots, a_{n-1\ 2}, a_{n\ 1}\}; \ldots ,$$

this is rewritten as[1]:

$$D_1 \equiv \{d_{1\ 1}\};$$
$$D_2 \equiv \{d_{2\ 1}, d_{2\ 2}\}; \ldots ; \qquad (5.2)$$
$$D_n \equiv \{d_{n\ 1}, d_{n\ 2}, \ldots, d_{n\ n-1}, d_{n\ n}\}$$

---

[1]The order in every diagonal is not important, but it is easy to define it so as to follow the same counting as Cantor.

The succession of these D_n defines the succession in Cantor's diagonal procedure:

$$a_{1\ 1}, a_{1\ 2}, a_{2\ 1}, \ldots, a_{1\ n}, a_{2\ n-1}, \ldots, a_{n-1\ 2}, a_{n\ 1}, \ldots. \tag{5.3}$$

But for

$$n \to \infty, \{D_n\} \leftrightarrow \aleph; \tag{5.4}$$

and

$$\begin{aligned}
D_n &\equiv \{a_{1\ n}, a_{2\ n-1}, \ldots, a_{n-1\ 2}, a_{n\ 1}\} \tag{5.5}\\
&\equiv \{d_{n\ 1}, d_{n\ 2}, \ldots, d_{n\ n-1}, d_{n\ n}\}\\
&\equiv \{d_m\}_n \leftrightarrow \aleph
\end{aligned}$$

for $m \in (1, n)$ and $n \to \infty$.

Therefore for $n \to \infty$ we have that $\{\{a_m\}_n\} \to \infty$ as $(\aleph \otimes \aleph)$. That is, as one proceeds along the diagonals these become greater and greater and, at the limit, infinite. Since also the number of the diagonals D_n tend to infinite at the limit (because we are speaking of limits), the rearrangement of Cantor does not offer any real advantage but to hide what was evident. In his demonstration, Cantor seems to forget the previous diagonals as he proceeds. But this results in simply demonstrating that, at the limit for infinite n, the diagonal D_n tends to "possess" infinite elements and that what is in one-to-one correlation with $\aleph$ is not the set of rational numbers, but only, at the limit, this last diagonal, ignoring all the previous diagonals which are, always at the limit, infinite in number[2]. Heuristically speaking, we have gone from:

$$
\begin{array}{ccccccc}
a_{11} & \rightarrow & a_{12} & \rightarrow & a_{13} & \rightarrow & a_{14} & \cdots & \rightarrow \infty \\
 & & & & & & & & \downarrow \\
a_{21} & \rightarrow & a_{22} & \rightarrow & a_{23} & \rightarrow & a_{24} & \cdots & \infty \\
\\
a_{31} & \rightarrow & a_{32} & \rightarrow & a_{33} & \rightarrow & a_{34} & \cdots \\
\\
a_{41} & \rightarrow & a_{42} & \rightarrow & a_{43} & \rightarrow & a_{44} & \cdots \\
\\
\cdots & & \cdots & & \cdots & & \cdots & \cdots
\end{array}
\tag{5.6}
$$

to:

$$
\begin{array}{ccccccc}
a_{11} & \rightarrow & a_{12} & & a_{13} & \rightarrow & a_{14} & \cdots & \infty \\
 & \swarrow & & \nearrow & & \swarrow & & & \nearrow \\
a_{21} & & a_{22} & & a_{23} & & a_{24} & \cdots & \searrow \\
\downarrow & \nearrow & & \swarrow & & & & & \infty \\
a_{31} & & a_{32} & & a_{33} & & a_{34} & \cdots \\
 & \swarrow & & & & & & \\
a_{41} & & a_{42} & & a_{43} & & a_{44} & \cdots \\
\\
\cdots & & \cdots & & \cdots & & \cdots & \cdots
\end{array}
\tag{5.7}
$$

Another way to represent the problem rigorously is the following: a bijective correspondence is represented mathematically by a function which can be inverted. In the present case, in order to represent the succession of denumerable "Cantorian" terms it is necessary to use a function to link the index of the succession, n, with the indices of the table, i and j. That is, we must be able to write $n = f(i, j)$. But this function, whatever it may be, cannot be inverted in that it is a function of two variables and to resolve it a further equation must be placed in system with the first, $n = g(i, j)$. From this we deduce that $f(i, j) = g(i, j)$,

---

[2]It is perhaps useful to remember that for $n \to \infty$, also $(n - k) \to \infty$, $\forall n, k \in \aleph$

which defines an implicit function of i and j, so that we will have $j = h(i)$ and $n = f(i, h(i))$. The condition necessary for n to be linked to i and j by a function which can be inverted is, therefore, that $j = h(i)$. This is exactly what happens when considering particular successions in the elements of the table, such as the diagonals, etc., as we will demonstrate more fully in the final section of this paper. If n is considered a constant, the equation $f(i, j) = n$ is an equation with two unknowns and, as above, its solution requires a further associated equation.

6 Natural numbers and denumerability

To structure the argument which follows in a rigorous manner, we must introduce the definition of a natural number. Subsequently it will be shown how it is possible to determine the non-denumerability of the rational numbers from the definition of natural numbers, not by means of graphic "demonstrations" but rather through logical and mathematical procedures. We begin by reporting the commonly accepted definition of a natural number given by G. Peano. In the following summary of the original version given in "Formulario Mathematico", we use modern notation rather than the original symbols.

Definition. Natural number:

$\aleph_0$ stands for *natural number.*

0 stands for *zero.*

$+$ stands for *plus.* If a is a natural number, $a+$ indicates the *number following* a.

Assume the three concepts $\aleph_0$, 0, $+$, to be primitive concepts by which we define every arithmetical symbol.

We determine the meaning of the undefined symbols $\aleph_0$, 0, $+$, by the following system of primitive propositions:

1. $\aleph_0$ is a class.

2. $0 \in \aleph_0$.

3. If $a \in \aleph_0$ then $a+ \in \aleph_0$.

4. If s is a class and $0 \in s$; and if $\forall a \in s$, $a+ \in s$; then $\aleph_0 \in s$.

5. If $a, b \in \aleph_0$ then $a+ = b+$ implies $a = b$.

6. If $a \in \aleph_0$ then $a+ \neq 0$.

This definition identifies the class of natural numbers and is characteristically linked to the principal of induction. It is however possible to make a generalization which will be very useful and extend the scope of the definition. This generalization is formed by substituting the symbol 0, linked to zero and therefore to a precise value, by a new symbol with a slightly more abstract meaning: O for origin, which is in any case equivalent to zero. Then, purely for convenience in subsequent steps, the symbol $+$ is substituted by $+_u$. This neither adds nor takes away anything from the original definition, we are simply introducing a symbol, u, which will simplify the identification of the symbol when it is correlated with real values. The new definition, which is reported below, has the advantage of being entirely abstract and of highlighting the fundamental structure of natural numbers; a definition of the concept of denumerability. Definition 2 denumerability: Let $\aleph$ be a class and let $+_u$ be a defined operation on elements of the class and call this operation "successive". Then, by definition: 1. $\exists O : O \in \aleph$. 2. $\forall A \in \aleph, a +_u \in \aleph$. 3. If $a, b \in \aleph$ then $a +_u = b +_u$ implies $a = b$. 4. If $a \in \aleph$ then $a +_u \neq O$. We will call $\aleph$ the class of denumerable numbers. We define "denumerable" every class isomorphic to $\aleph$, that is having the same structure as $\aleph$. For example, the natural numbers are denumerable in that it is enough to let $O = 0$ and $+_u = +1$ to obtain the usual definition of a natural number. If $O = 0$ and $+_u = +2$, we obtain the even natural numbers, while for $O = 1$ and $+_u = +2$ will have the odd natural number, etc. As we can see, denumerability clearly results from the structure of the definition of a class, as in the examples of classes of natural numbers, even natural numbers and odd natural numbers. That which defines a mathematical class is the property identified mathematically by an operator and the property of the natural numbers is identified by the operator of succession. The operator of succession is not, nor can it be defined, on the class of rational numbers globally. The rational numbers are defined by a double application of the denumerable, in particular of the whole numbers, of the form $(\aleph, \aleph)$, but perhaps it would be more explicit to write this in the form $\aleph(\aleph)$. It is not, therefore, equivalent to a definition of denumerable which uses

16

(applies) the definition of ℵ once only. Since the property of a class cannot be lost or changed if the representation is altered, this proof should be sufficient to dismiss Cantor's demonstration as erroneous.

7 The current definition of infinite and his paradox

At this point it only remains to clarify the thorny question posed by Cantor's diagonal representation: How is it possible that the rational numbers, and in particular the ordinate pairs, are not denumerable if it appears from the diagonal listing that all the elements of the table can be numbered? A first response was given in the first section, demonstrating that the succession defined by the diagonal procedure is not linear but divergent. A second reply comes from the mathematical definition of infinite. Until the 19th century, the definition of infinite was based, as it is based in this paper, on the principle of induction, that it on the definition of a natural number. Subsequently, in order to be able to justify the study of "actual" infinite numbers, a new definition was introduced, identifying the infinite set through the property of being able to be placed in bijection correspondence with a subset of itself. But it is clear that such a definition must be rejected in that it is self-contradicting: in fact to say that two sets are in bijection correspondence means that it is possible to put in relation each and every element of one set with each and every element of the other. If the second set is a proper subset of the first, only those elements belonging to the subset (or equivalent) can be "mapped" because of the principle of reflexivity of the relation of equivalence (generated by the bijective correspondence) and from the principle of identity. I.e there must be some elements of the superset not in bijection with the subset (for definition of subset itself). The above can be shown formally in the following manner: we consider the currently accepted definitions of proper subset, equivalent sets and infinite sets [The definition is taken from Kolmogorov - Fomin [4, §3. p.6]].

Definition 7.1. *Inclusion and proper inclusion. Let A and B be two sets. If all the elements constituting A also belong to B (not excluding the case A = B), then A is said to be a subset of B and is indicated by A ⊆ B. If A ≠ B and*

19

$A \neq \emptyset$ it is said to be a proper subset and is indicated by $A \subset B$.

Definition 7.2. *Equivalence of sets. Two sets M and N are said to be equivalent (notation $M \sim N$) if a biunivocal correspondence can be established between their elements.*

Definition 7.3. *Equipotency of sets. If two finite sets are equivalent, they are composed of the same number of elements. If instead two equivalent sets M and N are arbitrary, it is said that M and N have equal potency.*

Definition 7.4. *Biunivocal correspondence. The elements of two sets are said to be in biunivocal correspondence $(\rightleftharpoons)$ when each element in one set corresponds to one, and only one, element of the other and vice versa. It is evident that a biunivocal correspondence can be established between two finite sets if, and only if, they have the same number of elements .*

Definition 7.5. *Infinite set. Each infinite set is equivalent to a proper subset of itself. This property can be accepted as defining an infinite set.*

From definition [7.1], if $A \subset B$ then every element of A can be put in correspondence with one and only one element of B, $A \Rightarrow B$, but not the opposite $A \nRightarrow B$. Thus, rewriting the definition of an infinite set [7.5] using mathematical formalism, gives:

$$A, B \ infinites; \ A \subset B; \quad \text{by hypotesis} \tag{7.1}$$

$$A \sim B \Rightarrow A \rightleftharpoons B; \quad \text{by definition [7.2]} \tag{7.2}$$

$$A \subset B \Rightarrow A \rightharpoonup B, A \nrightarrow B; \quad \text{by definition [7.1]} \tag{7.3}$$

$$A \rightleftharpoons B, A \nrightleftharpoons; B \quad \text{from (7.2) and (7.3).} \tag{7.4}$$

We have, therefore, a contradiction in the currently accepted definition of an infinite set. The contradiction arises from the illusion that we can define an actual infinite and are able to manipulate it. A particular bijective correspondence is given by the identity function, I, which makes a given element of a set correspond to the element itself. The infinite is always in potency and never in act, at least in science. At the end, an answer of heuristic type can be given by one of Escher's drawings. Eyesight, being one of man's senses, is easily deceived and therefore, as the rationalists but also Aristotle affirmed, for problems of logic and mathematics we must scrupulously adhere to the strictest formal rigour. What emerges from the diagonal procedure is that, in effect, a form of extraction is performed to obtain a denumerable succession from a much richer whole, considering a single row or column of the doubly infinite table, and that the only real originality is that in this form we have managed to conceal its true nature.

8   Rational numbers and ordered pairs

To avoid any remaining doubts, due to the visual impact of Cantor's demonstration and not to its formal value, in this section we will show how the apparent existence of a successor in the table has a correspondent in the arithmetic representation and, vice versa, the property of non-existence of a successor to a rational number is also found in the tabular representation. In the usual representation of rational numbers, the demonstration of the non-existence of a successor is given by the theorem that between any two rational numbers it is always possible to find another rational number which is greater than the lower one and lesser than the higher one. We will show here how the same theorem is valid in the tabular representation of the rational numbers. If we consider the table ordered by rows, that is so that the second row is considered successive to the first and so on, we find, precisely because of the definition of a natural number and the construction of the table, that any element of the table has successors only in its own row. In other words, it is not possible to pass from one row to the successive applying the operator of succession. For example, no first element of a row has precedents, which is implicitly defined by the successor. To demonstrate the inverse we must first note how the principal difference between the two representations is the ordering. In the tabular representation, the order is imposed on the indices. The second index is ordered, in the usual way, by its entire value along a row; the first index does not vary along a row. It should be remembered that the second index is the denominator of the fraction representing the rational number. In the tabular representation, therefore, the order of the rational numbers along a row is completely different from that in the usual representation (hereafter called arithmetic representation). The first index is ordered down columns, while the second along rows; it describes the numerator of the fraction representing the rational number. Consider a generic term of the table (m, n), m, $n \in \aleph$. This term, together with all the terms of the general form (ml, nl), represents the rational number $\frac{m}{n}$. Consider, together with (ml, nl), $l \in \aleph$, the adjacent terms $(ml, nl + 1)$ which represent the rational number $\frac{m}{n}\left(1 - \frac{1}{nl+1}\right)$. When $l \to \infty$, the terms $\frac{m}{n}\left(1 - \frac{1}{nl+1}\right)$ tend to $\frac{m}{n}$ i.e. (m, n), $\forall m, n^1$ (see table 3.1, pag.5). Thus we see that the theorem discussed above is also valid in the tabular representation. In the proof given, the doubly infinite degree of freedom of the ordered pairs is in no way reduced. Many other interesting relations between the tabular and arithmetic representations of rational numbers can be found. Their interest lies in the possibility they offer to investigate the property of rational numbers from different (and equivalent) points of view.

---

[1]Other choices can be made, for example (ml, nl) and $(ml + 1, nl)$ that represents $\frac{m}{n}$ and $\frac{m}{n} + \frac{1}{nl}$ respectively.

9 Emerging properties

We show now that there is another case very similar to the construction of Rational Numbers from Natural Numbers. This case is well known by physicists: the "tensor product space". Naturally the "tensor product space" is much more complex then the Rational Numbers Set but there is a strong similarity in their construction. Let E_1, E_2 be linear spaces and $|\phi(1)>$, $|\chi(2)>$ two vectors of, respectively, E_1, E_2. If we call E the tensor product of E_1 and E_2: $E = E_1 \otimes E_2$, then to each pair of vectors $|\phi(1)>$, $|\chi(2)>$ of, respectively, E_1, E_2, corresponds a vector $|u_i(1)\rangle \otimes |v_l(2)\rangle$ of E. Well, it can be showed that there exist in E vectors that are not a tensor product of a vector in E_1 and a vector in E_2, the citation is taken from Cohen-Tannoudji, Diu, Laloe, (2, p.155):

(ii) There exist in E vectors which are not tensor products of a vector of E_1 by a vector of E_2. Since $|u_i(1)\rangle \otimes |v_l(2)\rangle$ constitutes by hypothesis a basis in E, the most general vector of E is expressed by:

$$|\psi\rangle = \sum_{i,l} c_{i,l} |u_i(1)\rangle \otimes |v_l(2)\rangle. \tag{9.1}$$

Given $N_1 N_2$ arbitrary complex numbers $c_{i,l}$, it is not always possible to put them in the form of products, $a_i b_l$, of N_1 numbers a_i and N_2 numbers b_l. Therefore, in general, vectors $|\phi(1)\rangle$ and $|\chi(2)\rangle$ of which $|\psi\rangle$ is the tensor product do not exist. However, an arbitrary vector of E can always be decomposed into a linear combination of tensor product vectors, as is shown in formula (F-7).

We can interpret this property saying that the space E is somehow greater then the product of the dimension of the single spaces E_1 and E_2. There are new properties that emerge from the composition of the two single spaces. We could say that this would have been predictable because reality wouldn't have been so complex as it is without this capacity or power. After all a furniture is not a mere wood pile. And would you be able to read the hour from some pieces of glass and steel? And, finally, think about the DNA.

10 Some Examples

The Dirichlet Function Measure

Let us consider the Dirichlet function:

$$d(x) = \begin{cases} 1, & for \, x = Rational \, Number \\ 0, & for \, x = Irrational \, Number \end{cases} \qquad (10.1)$$

This function is Lebesgue integrable in every set of points E. If E_R is the subset of E of the Rational Numbers and E_I the subset complement of E_R composed by the Irrational Numbers, its Lebesgue integral is (considering the Rational Numbers Set denumerable):

$$\int_E d(x)dx = \int_{E_R} d(x)dx + \int_{E_I} d(x)dx = \int_{E_R} d(x)dx = \mu(E_R) = 0 \qquad (10.2)$$

because $d(x) \equiv 0$ for $x \in E_I$ and $\mu(E_R) = 0$ for every set of Natural Numbers is denumerable and so its Lebesgue measure is zero. But considering the Rational Numbers Set not denumerable, $\mu(E_R) \neq 0$ and so even the Lebesgue integral of d(x) is not null. This is, in my opinion, reasonable because the Rational Numbers Set is dense in the Real Number Set so that in every neighborhood of a Real Number (and so also of every Irrational Number) there are infinite Rational Numbers. It appears to me few understandable that all the measure of a Real Numbers Set lies only on the Irrational Numbers Subset.

Not Equivalence of Power and Trigonometric Sets in L^2

Among the complete set in L^2 there are the Trigonometric Function Set:

$$1, sin(x), cos(x), sin(2x), cos(2x), ; x \in [0, 2\pi); \qquad (10.3)$$

$$with\ Weight\ Function\ p(x) = 1;$$

written synthetically as:

$$\{sin(mx), cos(nx)\}; m, n \in \aleph_0, x \in [0, 2\pi); \tag{10.4}$$
$$with\ Weight\ Function\ p(x) = 1;$$

and the Power Set:

$$1, x, x_2, ;\ x \in \forall(a, b), a, b \in \Re; \tag{10.5}$$
$$valid\ with\ every\ Weight\ Function;$$

written synthetically as:

$$\{x^n\};\ n \in \aleph_0,\ x \in \forall(a, b), a, b \in \Re; \tag{10.6}$$
$$valid\ with\ every\ Weight\ Function.$$

The first is an $(\aleph, \aleph)$ Set while the second is an $(\aleph)$ Set. Actually they are considered equivalent because the first has the power of Rational Numbers Set that is equal to the power of Natural Numbers Set, given that Rational Numbers Set is considered denumerable. We will verify hereafter if this is the case. Let us consider the two Complete Sets above in the same range:

$$\{sin(mx), cos(nx)\}; m, n \in \aleph_0, x \in [0, 2\pi); \tag{10.7}$$
$$with\ Weight\ Function\ p(x) = 1;$$
$$\{x^n\};\ n \in \aleph_0,\ x \in [0, 2\pi); \tag{10.8}$$
$$with\ Weight\ Function\ p(x) = 1.$$

If both of them are complete in the same range, every function f in $L^2_{[0,2\pi]}$ can be expressed as a sum of functions of the two sets:

$$f_\phi = \sum_{i=0}^{\infty} a_i x^i; \ i \in \aleph, \ x \in [0, 2\pi], a_i \in C; \qquad (10.9)$$

$$f_\psi = \sum_{m,n=0}^{\infty} (b_m sin(mx) + c_n cos(nx)); \qquad (10.10)$$

$$m, n \in \aleph, \ x \in [0, 2\pi], b_m, c_n \in C;$$

The subscripts ϕ, ψ are for subsequent use, as we'll see. Let us call $|\phi_i >$ the vector related to x^i of the base $\{x^i\}$ and $|\psi_{1,m} >$ the vector related to $sin(mx)$, $|\psi_{2,n} >$ the vector related to $cos(nx)$ (we use two indices m and n to remark that they are indices independent).

Let us start with f_ϕ let us project it on the ψ basis:

$$\sum_{m,n=0}^{\infty} (|\psi_{1,m} >< \psi_{1,m}| + |\psi_{2,n} >< \psi_{2,n}|) \, |f_\phi >= \qquad (10.11)$$

$$= \sum_{m,n=0}^{\infty} (|\psi_{1,m} >< \psi_{1,m}||f_\phi > +|\psi_{2,n} >< \psi_{2,n}|f_\phi >) =$$

$$= \sum_{m,n=0}^{\infty} \left(|\psi_{1,m} > \sum_{i=0}^{\infty} a_i < \psi_{1,m}|\phi_i > + \right.$$

$$\left. +|\psi_{2,n} > \sum_{i=0}^{\infty} a_i < \psi_{2,n}||\phi_1 > \right) =$$

$$= \sum_{m,n,i=0}^{\infty} a_i \, (< \psi_{1,m}|\phi_i > |\psi_{1,m} > + < \psi_{2,n}||\phi_1 > |\psi_{2,n} >)$$

This is a vector in the $\{\psi\}$ space:

$$f_\psi = \sum_{m,n=0}^{\infty} (b_m|\psi_{1,m} > +c_n|\psi_{2,n} >); \qquad (10.12)$$

with:

$$\begin{cases} b_m & = & \sum_{i=0}^{\infty} a_i < \psi_{1,m}|\phi_i > \\ \\ c_n & = & \sum_{i=0}^{\infty} a_i < \psi_{2,n}|\phi_i > \end{cases} \qquad (10.13)$$

that can be written also as:

$$\begin{cases} b_m & = & < \psi_{1,m}| \sum_{i=0}^{\infty} a_i\phi_i > \\ \\ c_n & = & < \psi_{2,n}| \sum_{i=0}^{\infty} a_i\phi_i > \end{cases} \qquad (10.14)$$

For each b_m and c_n there is one and only one $< \psi_{1,m}| \sum_{i=0}^{\infty} a_i\phi_i >$ and $< \psi_{2,n}| \sum_{i=0}^{\infty} a_i\phi_i >$. Now let us project f_ψ on the ϕ basis:

$$\sum_{i=0}^{\infty} (|\phi_i >< \phi_i|) |f_\psi >= \qquad (10.15)$$

$$= \sum_{i=0}^{\infty} (|\phi_i >< \phi_i|) \left(\sum_{m,n=0}^{\infty} (b_m|\psi_{1,m} > +c_n|\psi_{2,n} >) \right) =$$

$$= \sum_{i=0}^{\infty} |\phi_i > \left(\sum_{m=0}^{\infty} b_m < \phi_i|\psi_{1,m} > + \sum_{n=0}^{\infty} c_n < \phi_i|\psi_{2,n} > \right)$$

This is a vector in the $\{\phi\}$ space:

$$f_\phi = \sum_{i=0}^{\infty} a_i|\phi_i >; \qquad (10.16)$$

with:

$$a_i = \sum_{m=0}^{\infty} b_m < \phi_i|\psi_{1,m} > + \sum_{n=0}^{\infty} c_n < \phi_i|\psi_{2,n} > \qquad (10.17)$$

that can be written also as:

$$a_i =< \phi_i| \sum_{m=0}^{\infty} b_m\psi_{1,m} > + < \phi_i| \sum_{n=0}^{\infty} c_n\psi_{2,n} > \qquad (10.18)$$

Now, differently from above, we have an infinite choice of elements $< \phi_i| \sum_{m=0}^{\infty} b_m \psi_{1,m} >$ or $< \phi_i| \sum_{n=0}^{\infty} c_n \psi_{2,n} >$! For example we can choose, for each a_i, one $< \phi_i| \sum_{m=0}^{\infty} b_m \psi_{1,m} >$ and then we'll have:

$$< \phi_i| \sum_{n=0}^{\infty} c_n \psi_{2,n} >= a_i - < \phi_i| \sum_{m=0}^{\infty} b_m \psi_{1,m} > . \qquad (10.19)$$

There are infinite vectors of the ψ space that have the same projection on the ϕ space.

11 Conclusions

The nineteenth century saw a huge enhancement of the research in all the fields of science, led by the the industrial revolution and its needs. The great discoveries gave to the humankind a sense of power never felt before. In that atmosphere scientists threw themselves enthusiastically in the research of the very origin of "all". Cosmology on one side, atomic and nuclear physics on the other side, received a great boost in the physical realm. In the mathematical realm, among the others, the research of an axiomatization and formalization of the mathematics were two of the main fields. In that context the studies of Sets led to the (ancient) question of the Infinite Sets and their definitions and properties. In my opinion it was somewhat a delirium of power that led to believe that the human beings could reach, by means of science, the complete knowledge and domination of the Universe. Laically speaking this is a huge logical mistake, in my opinion, because the content never can contain its container (I'm sorry for the cacophony). We humans are part of the Universe so we cannot comprehend it all (note the double meaning)! But the previous statement has not to be taken as diminutive of the potentials of humankind, at the contrary! We experience every day the dichotomy of our nature: physical and mental. Our mind is capable of abstractions and of discoveries that enrich our Universe. They resemble the somewhat mysterious "emerging properties". Maybe every time we make a discovery we enrich the Universe expanding it and at every expansion there are more and more things to discover. In a sense we are (maybe!) participating to the creation of our Universe. I want to stress that I'm talking here from a laic, a philosophical point of view, theology is beyond the scope of these reasonings.

What is the Universe, where does it come from are questions that own to philosophy and theology. In particular philosophy should take back its prerogative to be the bridge between science and theology.

In the end I want to thank all of you readers. I'm conscious that this book

33

has many lacks and I apologize for them, I made my best.

Roberto Musmeci

Bibliography

[1] G. Cantor. Über eine eigenschaft des inbegriffes aller reellen algebraischen zahlen. *Journal für reine und angewandte Mathematik*, 77:258–62, 1874.

[2] Cohen-Tannoudji. Bernard Diu. Franck Laloë. *Quantum Mechanics*. Wiley-Interscience publication, 1977.

[3] A. N. Kolmogorov. S. V. Fomin. *Elements of the Theory of Function and Functional Analysis*. Dover Publication, Inc., Mineola, New York, 1999.

[4] G. Peano. *Formulario Mathematico*. Edizioni Cremonese, Roma, 1960.

www.ingramcontent.com/pod-product-compliance
Lightning Source LLC
Chambersburg PA
CBHW021850170526
45157CB00006B/2384